ANONYMOUS PROXY SERVER

These computers have applications that download webpages without revealing the user's identity. They help bad guys hide while traveling and launching attacks in cyberspace.

01000011
01111001
01101110
01101010
01100001

BINARY CODE

Computers speak in code. Binary code is based on two digits, 0 or 1. By creating different combinations of these two digits, programmers tell a computer what to do.

CIPHER

Secrets are shared in cyberspace by encrypting information. But beware; some bad guys can break your secret code.

DARKNET

Cyber criminals often use a private network to hide their illegal activities.

DNS SINKHOLE

A sinkhole helps stop botnets. Good guys disguise a DNS server as a target. Zombies attack the decoy instead because they don't know any better.

FOR RODNEY JOFFE.

YOUR PASSION FOR CYBERSECURITY IS WHAT BROUGHT US TOGETHER TO WRITE THIS BOOK. YOU'RE THE NOBLE WARRIOR WE HOPE OUR NEXT GENERATION WILL BECOME.

- CHASE & HEATHER

FOR MY WONDERFUL LITTLE GIRLS, CALLIE AND CAELYN. LEARN SOMETHING EVERYDAY AND LOOK FOR THE MAGIC IN EVERYTHING. DADDY LOVES YOU.

- CHASE

FOR THE SPECIAL YOUNG CYBER WARRIORS IN MY LIFE WHO ARE MY INSPIRATION FOR THIS BOOK, GRANT, HANNAH, OWEN AND HENRY.

- HEATHER

FOR MY WONDERFUL WIFE STARLA WITHOUT WHOM THIS WOULD HAVE BEEN IMPOSSIBLE AND MY TWO CYNJETTES SAKURA AND HINATA.

- SHIROW

OUR BATTLES ARE JUST BEGINNING!

BE THE CYNJA!!
INFOSEC RULES!
♡ Heather

CYBERSPACE NEEDS U!

THE CYNJA

VOLUME 1

STORY BY

CHASE CUNNINGHAM
&
HEATHER C. DAHL

ILLUSTRATED BY

SHIROW DI ROSSO

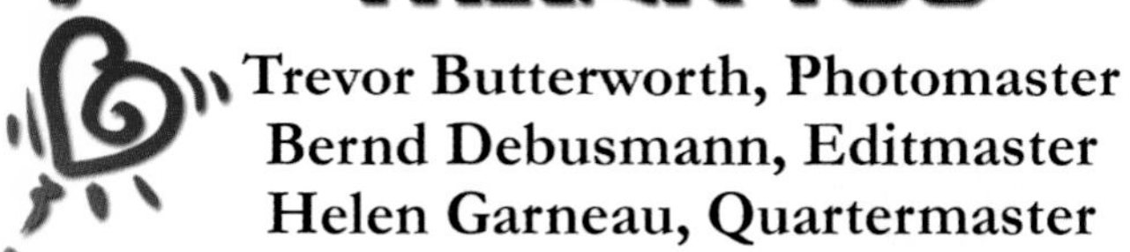

THANK YOU

Trevor Butterworth, Photomaster
Bernd Debusmann, Editmaster
Helen Garneau, Quartermaster

All illustrations by
www.mentalpictures.be

For information about special discounts for bulk purchases or speaking events, please contact info@thecynja.com

ISBN 978-0-9912049-0-8 (Hardcover)

Published by Blurb
First Edition

Publisher's Cataloging-in-Publication
(Provided by Quality Books, Inc.)

Cunningham, Chase.
The Cynja. Volume 1, Our battles are just beginning!
story by Chase Cunningham & Heather C. Dahl ;
illustrated by Shirow Di Rosso.
pages cm
SUMMARY: Deep inside the planet's virtual world, lurking in the darkened cyber alleys of digital neighborhoods, zombies, worms, and botnets all threaten the cybercitizens' happiness and future.
Audience: Ages 6-12.
ISBN 978-0-9912049-0-8
ISBN 978-0-9912049-1-5
ISBN 978-0-9912049-2-2
ISBN 978-0-9912049-3-9

1. Internet--Juvenile fiction. 2. Computer security --Juvenile fiction. 3. Science fiction.
1. Internet-- Fiction. 2. Computer security--Fiction.
3. Computers-- Fiction. 4. Science fiction.
I. Dahl, Heather C. II. Di Rosso, Shirow, illustrator.
III. Title. IV. Title: Our battles are just beginning!

PZ7.C916285Cy 2014 [E]
QBI13-2547

Printed in China

THIS IS THE WORLD WE ALL KNOW AND LOVE,
BUT THERE IS A DARKER SIDE TO IT ALL....
DIGITAL CHAOS IS WREAKING HAVOC ON OUR LIVES.
IT'S AN ERA WHEN CRIMINALS ARE HIDDEN DEEP INSIDE
THE PLANET'S CYBER ALLEYS AND NEIGHBORHOODS.
OUR HAPPINESS AND FUTURE ARE BEING THREATENED.
THE BAD GUYS ARE ONLY GETTING WORSE AND WORSE.
THEY'RE SUBVERTING OUR LIVES AND INFECTING OUR
COMPUTERS. THESE CYBER OUTLAWS ARE INVISIBLE TO
ALL BUT A FEW. ONLY THOSE TRAINED
IN THE POWERS OF
CYBERSPACE CAN SEE
THESE ENEMIES.

THE WORST – THE MOST WICKED – OF THESE CRIMINALS IS THE BOTMASTER. HIS GOAL IS TO INFECT MILLIONS OF COMPUTER NETWORKS AND GAIN CONTROL OF THE WORLD. HIS WEAPON IS A BINARY POWER STAFF, WHICH HE USES TO HACK PEOPLE'S COMPUTERS.

ONCE THE BOTMASTER HAS HACKED A WEAK COMPUTER, IT BECOMES A ZOMBIE – AND IS UNDER HIS CONTROL. AS HE HACKS COMPUTER AFTER COMPUTER, HIS ZOMBIE ARMY GROWS, STEALTHILY, INTO A BOTNET. ONCE THESE NASTY BOTS ARE STAGED ALONG THE DOMAIN NAME SYSTEM THEY ARE READY TO ATTACK,

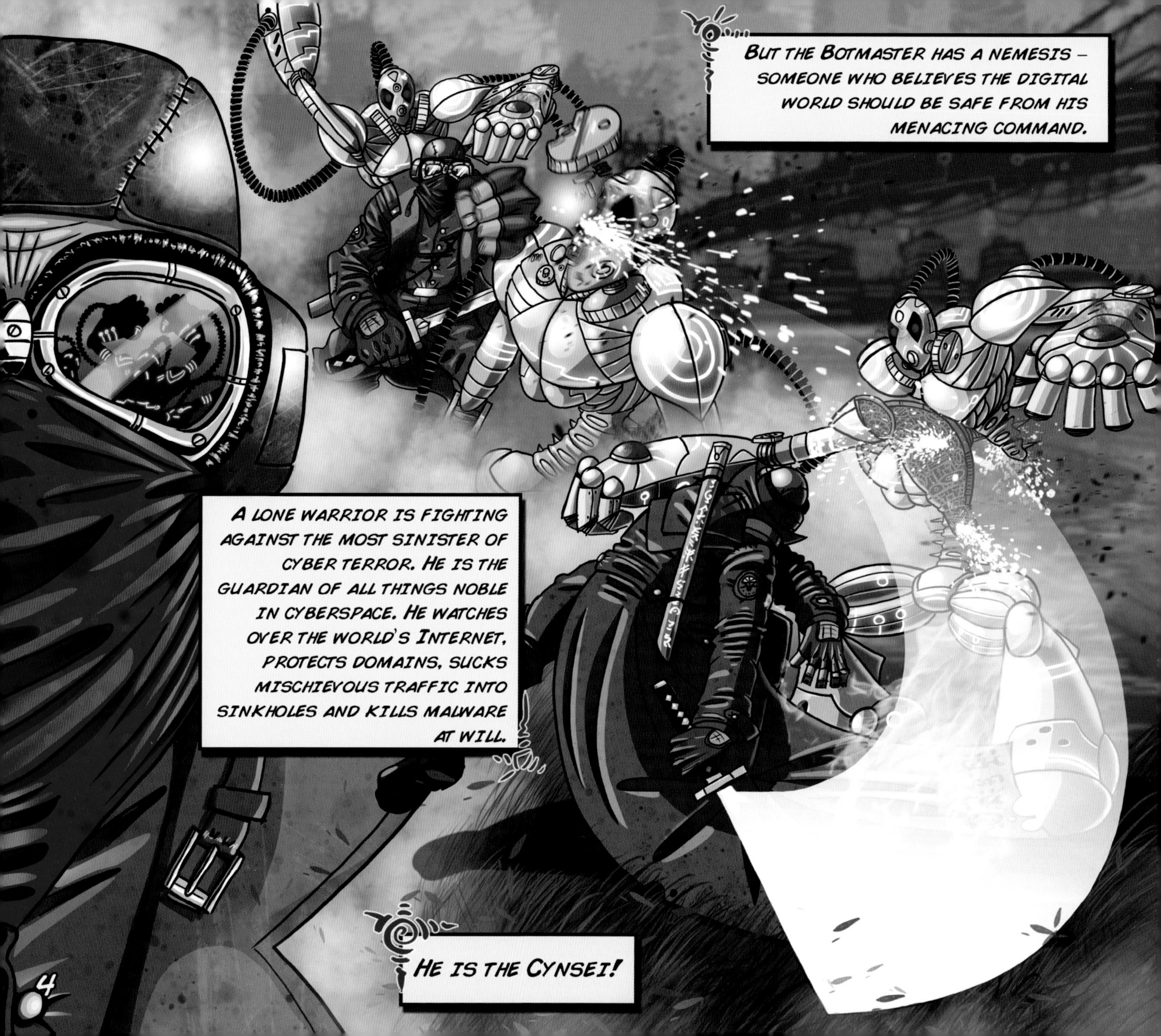
BUT THE BOTMASTER HAS A NEMESIS – SOMEONE WHO BELIEVES THE DIGITAL WORLD SHOULD BE SAFE FROM HIS MENACING COMMAND.
A LONE WARRIOR IS FIGHTING AGAINST THE MOST SINISTER OF CYBER TERROR. HE IS THE GUARDIAN OF ALL THINGS NOBLE IN CYBERSPACE. HE WATCHES OVER THE WORLD'S INTERNET, PROTECTS DOMAINS, SUCKS MISCHIEVOUS TRAFFIC INTO SINKHOLES AND KILLS MALWARE AT WILL.
HE IS THE CYNSEI!

BIP!
SUDDENLY MALWARE BEGINS SCROLLING ACROSS THE INTERNET LIKE AN ANGRY MOB.
THE CYNSEI REACHES FOR HIS SECURITY MONITOR AND SEES THE BOTMASTER IS BACK TO FIGHT AGAIN.
UGH!
SMASH!
THE BOTMASTER IS A RELENTLESS FOE, DEVIOUS, MORE CLEVER THAN ANY CRIMINAL IN HISTORY. AND HE IS ABOUT TO UNLEASH THE MOST PERILOUS MALWARE OF OUR GENERATION, ZEUS.
IT WILL INFECT COMPUTERS AND CREATE ZOMBIES FASTER THAN THE CYNSEI CAN FIGHT.
IF THE BOTMASTER WINS, PEOPLE AROUND THE WORLD WILL LOSE THEIR MONEY, THEIR DATA AND POSSIBLY THEIR LIVES!

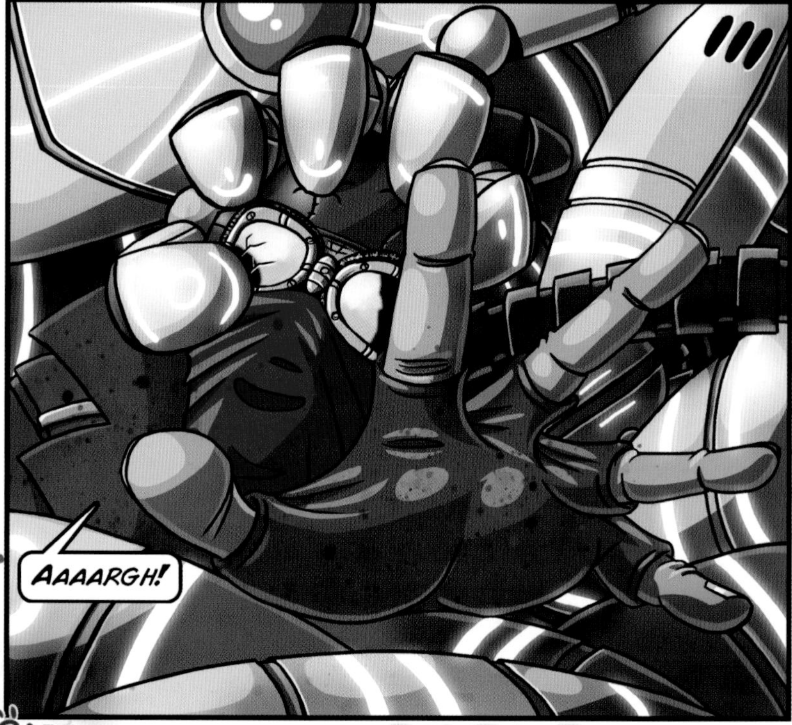

THE INTERNET IS AT RISK. AS SOON AS THE CYNSEI FINDS WHERE A BAD GUY HAS INFILTRATED A PROTECTED DOMAIN, ANOTHER BAD GUY ATTACKS FROM A DIFFERENT VECTOR.

HE KEEPS PLUGGING HOLES.
A FEW YEARS AGO THE CYNSEI WAS WINNING, BUT NOW NEW SECRET TROJANS, INFECTIOUS WORMS AND MENACING BOTNETS ARE GAINING SPEED.

And the Cynsei knows that in order to protect all good people against digital enemies he needs help. But who would be brave enough to travel into the depths of malicious networks? A special cyber warrior who will demolish any evil lurking inside our computers.

A CYNJA, that's who he needs and he needs one fast.

JEFFERSON MIDDLE SCHOOL

THIS IS GRANT WILEY AND HE LIKES PLAYING VIDEO GAMES A LOT.
IN FACT, HE LIKES BUILDING VIDEO GAMES EVEN MORE. CREATING SOMETHING OUT OF NOTHING MAKES GRANT FEEL LIKE A WIZARD IN HIS OWN DIGITAL WORLD.

WHEN GRANT TURNED 11 YEARS OLD HIS PARENTS GAVE HIM A SPECIAL BIRTHDAY GIFT, HIS VERY OWN COMPUTER.

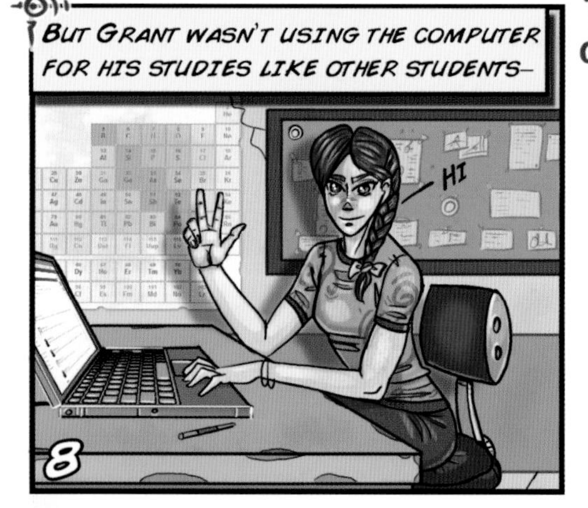
BUT GRANT WASN'T USING THE COMPUTER FOR HIS STUDIES LIKE OTHER STUDENTS–
HI

INSTEAD HE WAS SECRETLY SCRIPTING PROGRAMS.
HEY

And that's why Grant had to stay after school on this fateful day.

Grant was programming a new game instead of finishing his work.

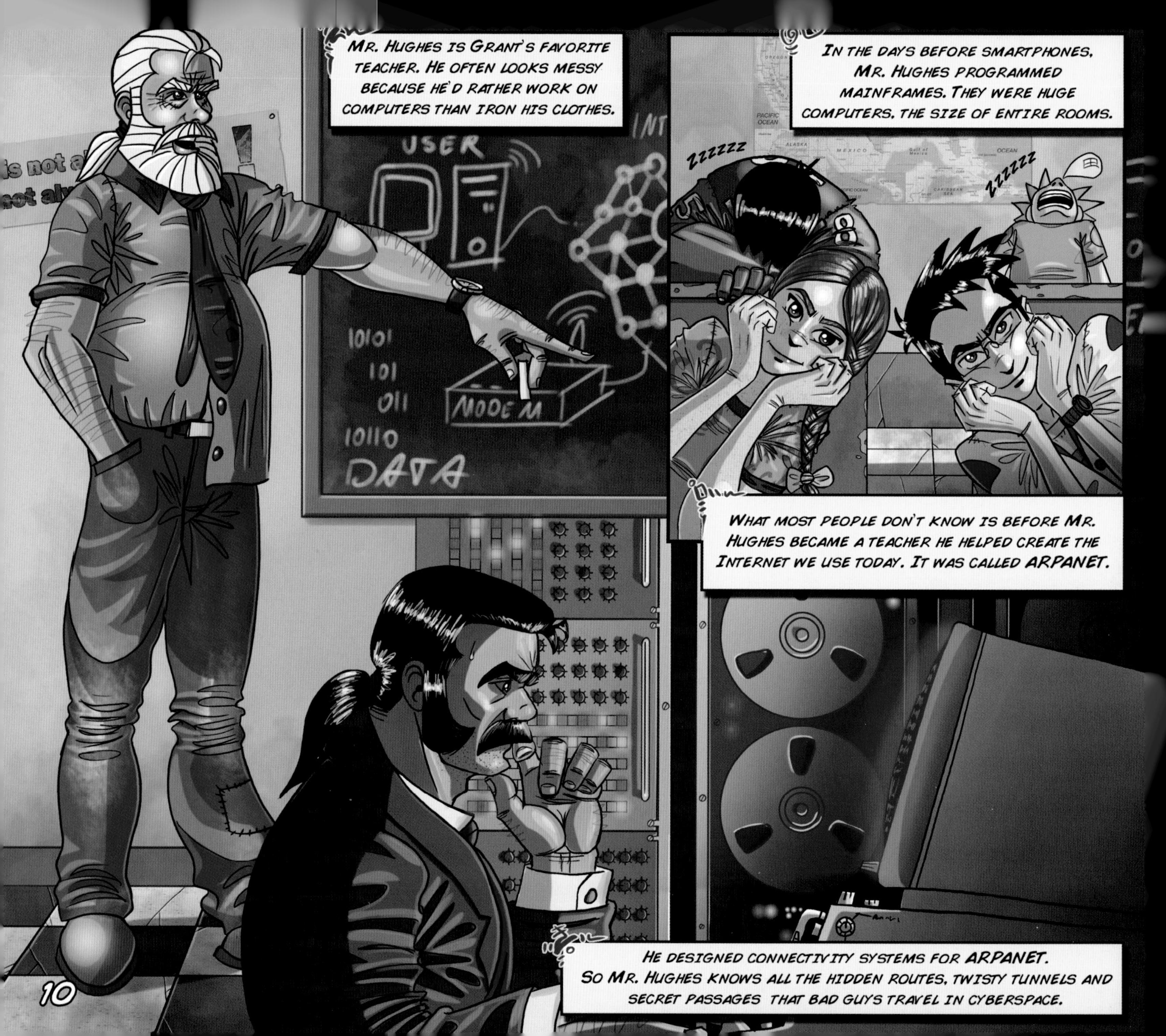
MR. HUGHES IS GRANT'S FAVORITE TEACHER. HE OFTEN LOOKS MESSY BECAUSE HE'D RATHER WORK ON COMPUTERS THAN IRON HIS CLOTHES.
USER
10101
101
011
10110
MODEM
DATA
IN THE DAYS BEFORE SMARTPHONES, MR. HUGHES PROGRAMMED MAINFRAMES. THEY WERE HUGE COMPUTERS, THE SIZE OF ENTIRE ROOMS.
ZZZZZZ
ZZZZZZ
WHAT MOST PEOPLE DON'T KNOW IS BEFORE MR. HUGHES BECAME A TEACHER HE HELPED CREATE THE INTERNET WE USE TODAY. IT WAS CALLED ARPANET.
HE DESIGNED CONNECTIVITY SYSTEMS FOR ARPANET.
SO MR. HUGHES KNOWS ALL THE HIDDEN ROUTES, TWISTY TUNNELS AND SECRET PASSAGES THAT BAD GUYS TRAVEL IN CYBERSPACE.

MEANWHILE GRANT IS TRYING TO FINISH HIS HOMEWORK.
TIC
TIC

TIC
TIC

TICITICITI
ICITICITI
TIC
BUT THE SOUND OF MR. HUGHES TYPING LIGHTNING FAST IS DISTRACTING.

TICITICITICITICITICITIC
ICITICITICITICITICITICITI

TICITICITICITICITICITI

BOOM!

FSSSSSHH
WITHIN A NANOSECOND, MR. HUGHES VANISHED. LEAVING NOTHING BUT A SMELL OF TOXIC SMOKE AND A BURN MARK.

PLUGGED INTO THE PORTAL
WAS A USB STICK UNLIKE ANY
HE'D EVER SEEN.
IT SEEMED TO BE MAGIC.

KNOCK
KNOCK
KNOCK

GRANT, HONEY, LET'S GO HOME.

OK, MOM!
WHATEVER THIS IS, IT MUST BE IMPORTANT. I THINK MR. HUGHES NEEDS MY HELP.

MOM, SOMETHING WEIRD HAPPENED!
WHAT, DEAR?
UH, NEVER MIND.

THE WILEY HOME
31

WHERE IS IT?
AHA!

LET'S SEE WHAT THIS THING DOES.

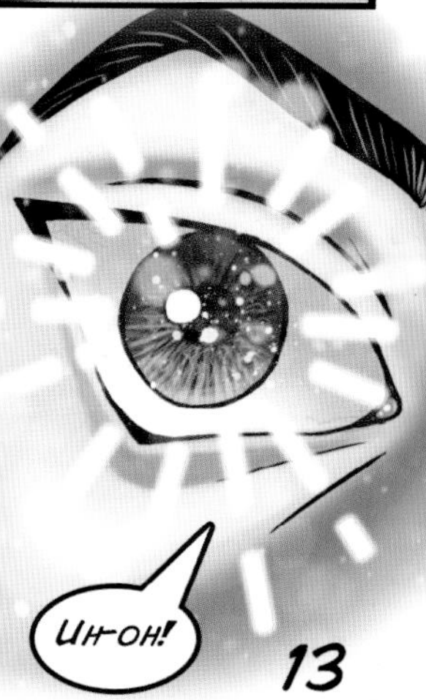
UH-OH!

GRANT WAS SUCKED INTO THE INTERNET. FLYING ALONGSIDE TRILLIONS OF DATA PACKETS, MOVING AT THE SPEED OF 3600 KILOBITS PER SECOND,
AAAAAAAAAAAH!!!
HE WAS BOUNCED OVER AN ENCRYPTED CHANNEL AND ROUTED INTO OBLIVION.

OOOOOH MYYYYY GOSH!!

GRANT TUMBLED THROUGH A NEWLY OPENED PORT INTO A MASSIVE EXPLODING FIREWALL.
WHOOSH!
AWW!
THUD!

MAN. THIS HURTS!

WHERE AM I?

WOW!

WHAT??

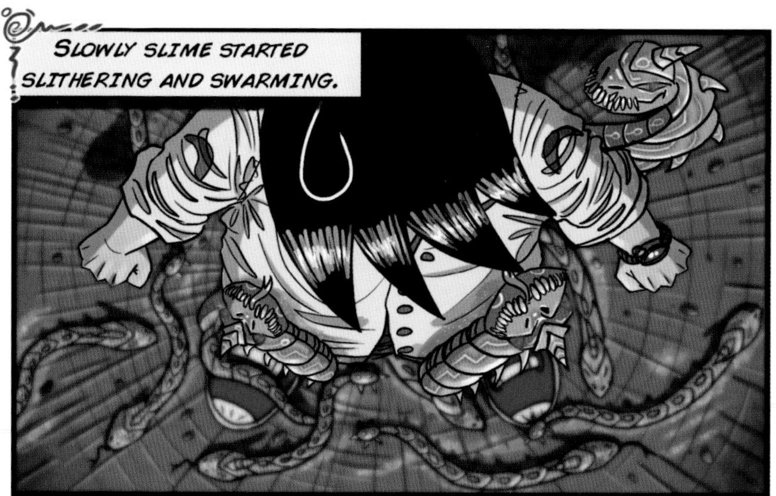
SLOWLY SLIME STARTED SLITHERING AND SWARMING.

AAAAAAH!
IT WAS POURING FROM HOLES IN THE GROUND, SQUISHED OUT BY DISGUSTING DIGITAL WORMS.

AAAAAAH!
HUNDREDS OF WORMS...

AAAAAAAH!
THESE MALICIOUS PARASITES OF THE INTERNET INFECT HEALTHY SYSTEMS WITH SICKLY CODE.

PLEASE SOMEONE HELP ME!!

GRAB
FLASH

YANK

WHAT NOW?

I AM THE CYNSEI!

MY WHAT?
WHAT THE HECK IS GOING ON? WHERE AM I?
I WANT TO GO HOME.
NOW!
I WAS JUST CHECKING OUT THIS CRAZY USB AND THEN...

SILENCE, YOUNG GRANT. YOUR DESTINY IS TO PROTECT THE WORLD FROM EVIL.
YOU HAVE EXCELLED AT COMPUTER PROGRAMMING.

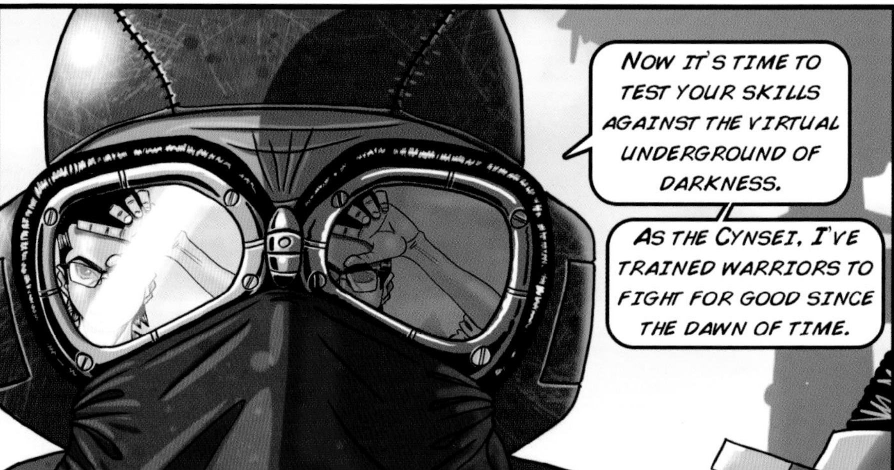
NOW IT'S TIME TO TEST YOUR SKILLS AGAINST THE VIRTUAL UNDERGROUND OF DARKNESS.
AS THE CYNSEI, I'VE TRAINED WARRIORS TO FIGHT FOR GOOD SINCE THE DAWN OF TIME.

"FIRST AS AN EGYPTIAN PHARAOH, I WAS GIVEN IMMORTALITY BY THE ANCIENT GODS AS LONG AS I PROTECTED FUTURE CIVILIZATIONS FROM THE WRATH OF EVIL."
"IN THE MIDDLE AGES, A KING GRANTED ME KNIGHTHOOD SO I COULD PROTECT OUR CODE OF HONOR AGAINST BARBARIANS."
"WORLD WAR I INTRODUCED A NEW BATTLEGROUND AND A NEW WARRIOR. I TOOK TO THE AIR DOGFIGHTING ENEMIES IN THE SKY."
"EVENTUALLY I BECAME A SAMURAI AND UNLEASHED MY DEADLY MARTIAL ARTS ON THE ENEMY."
"HUNDREDS OF YEARS LATER THE WORLD CHANGED AGAIN. THIS TIME OUTLAWS ROAMED THE WILD WEST. I WAS A COWBOY SEEKING JUSTICE FOR THOSE WHO KNEW NONE."

"GRANT, TODAY OUR BATTLEFIELD IS A NEW FRONTIER."

"IT'S ONE THAT'S INVISIBLE TO ALL BUT THOSE WITH CYBER POWERS."

"IT'S TIME FOR ME TO TRAIN A NEW TYPE OF WARRIOR, ONE WHO WILL PROTECT CYBERSPACE LIKE THOSE WHO PROTECTED MANKIND THROUGHOUT HISTORY."

TODAY, INSIDE MY PROTECTED NETWORK YOUR TRAINING WILL BEGIN.
GRANT, THE WORLD NEEDS BOYS AND GIRLS WITH YOUR DIGITAL SKILLS TO BECOME CYNJAS AND KEEP THE INTERNET SAFE FOR ALL.
WOAH, MR. HUGHES? YOU'RE THE CYNSEI!!!

WHAT'S A CYNJA?

A CYNJA IS A CYBER NINJA.
THEY FIGHT MALICIOUS OPERATORS.
THESE RUTHLESS VILLAINS ATTACK OUR COMPUTERS, THREATEN OUR FAMILIES, AND RUIN ALL THAT IS GOOD IN CYBERSPACE.

THE BAD GUYS STEAL MILLIONS AND MILLIONS OF DOLLARS.
THEIR DIGITAL CRIMES CAUSE GREAT DAMAGE. ALL OF US ARE AT RISK.

YOU MEAN BAD GUYS LIVE IN OUR COMPUTERS?

YES, CYNJA.
AT THIS VERY MOMENT, THEY'RE PLOTTING THEIR NEXT ATTACK. I NEED YOU TO HELP ME.

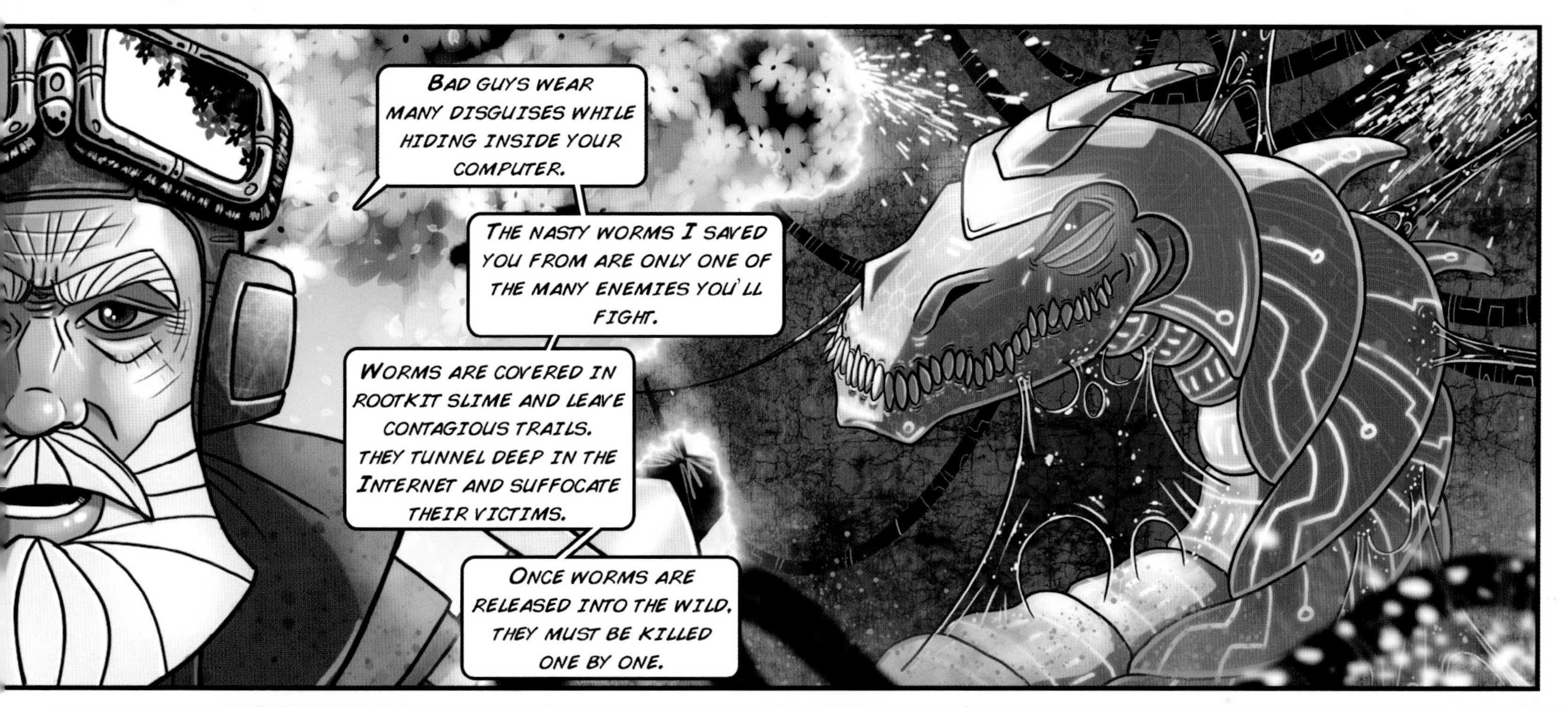
BAD GUYS WEAR MANY DISGUISES WHILE HIDING INSIDE YOUR COMPUTER.
THE NASTY WORMS I SAVED YOU FROM ARE ONLY ONE OF THE MANY ENEMIES YOU'LL FIGHT.
WORMS ARE COVERED IN ROOTKIT SLIME AND LEAVE CONTAGIOUS TRAILS. THEY TUNNEL DEEP IN THE INTERNET AND SUFFOCATE THEIR VICTIMS.
ONCE WORMS ARE RELEASED INTO THE WILD, THEY MUST BE KILLED ONE BY ONE.

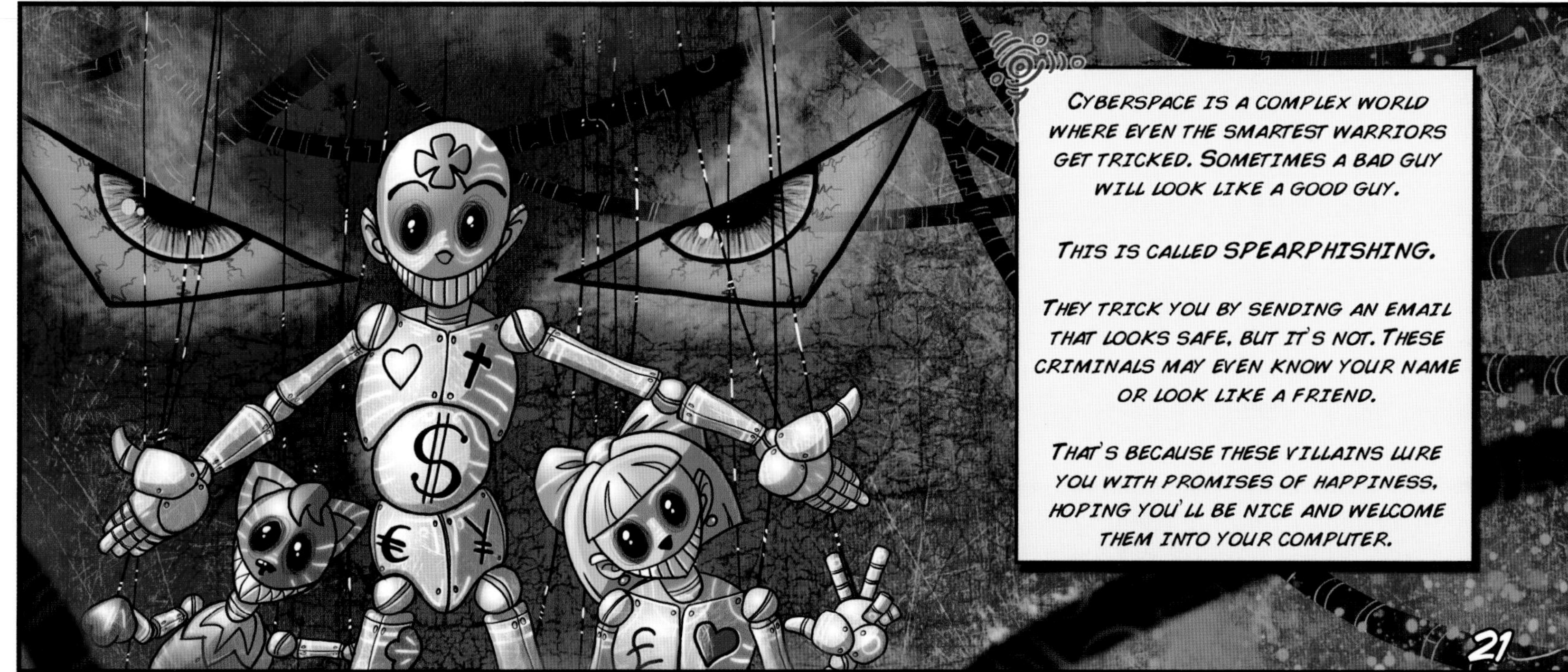
CYBERSPACE IS A COMPLEX WORLD WHERE EVEN THE SMARTEST WARRIORS GET TRICKED. SOMETIMES A BAD GUY WILL LOOK LIKE A GOOD GUY.
THIS IS CALLED SPEARPHISHING.
THEY TRICK YOU BY SENDING AN EMAIL THAT LOOKS SAFE, BUT IT'S NOT. THESE CRIMINALS MAY EVEN KNOW YOUR NAME OR LOOK LIKE A FRIEND.
THAT'S BECAUSE THESE VILLAINS LURE YOU WITH PROMISES OF HAPPINESS, HOPING YOU'LL BE NICE AND WELCOME THEM INTO YOUR COMPUTER.

ANOTHER ENEMY WE MUST BATTLE IS THE DEVIOUS TROJAN.
THEY APPEAR HARMLESS BECAUSE THEY LOOK NICE. SOME WANT YOU TO BELIEVE THEY'LL MAKE YOUR OPERATING SYSTEM WORK BETTER. BUT DON'T FALL FOR THEIR DECEPTION. BEHIND THEIR PRETTY FACES, TROJANS ARE UGLY.
THEY GIVE BAD HACKERS REMOTE CONTROL OF YOUR COMPUTER TO MAKE MISCHIEF. IT'S POSSIBLE TROJAN MALWARE MIGHT STEAL YOUR FAMILY'S MONEY OR EVEN DESTROY YOUR ENTIRE COMPUTER.

YOUNG CYNJA...
... MY THREAT SENSORS SHOW THE BOTMASTER IS LAUNCHING A MASSIVE DENIAL-OF-SERVICE ATTACK.
HE'S CREATING A BOTNET SO BIG IT WILL FLOOD THE WORLD'S SERVERS WITH MILLIONS OF QUERIES PER SECOND.
WEBSITES WILL SLOW UNTIL THEY STOP. THE ENSUING CHAOS WILL CREATE COVER FOR BAD GUYS TO STEAL A MASSIVE SUM OF MONEY.

YOU'VE DEMONSTRATED THE SKILLS TO BECOME A BETA CYNJA.
BUT YOU'RE STILL YOUNG AND HAVE MANY TESTS AHEAD.
ONLY THE MOST AGILE, SMART AND STRONG WILL SURVIVE THIS TRAINING.

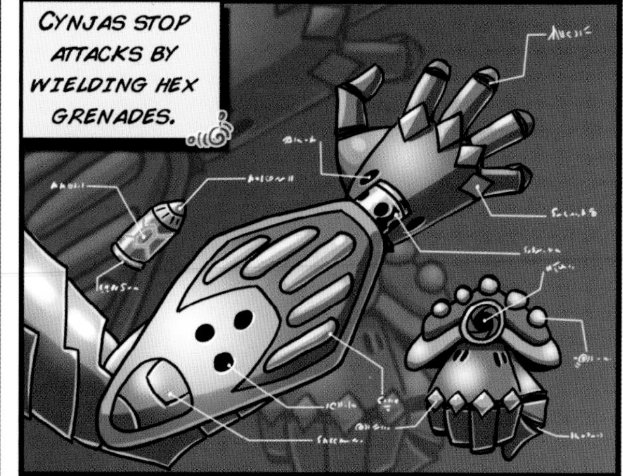
CYNJAS STOP ATTACKS BY WIELDING HEX GRENADES.

THEY WEAR A POWERFUL BINARY VISION MONOCLE FOR DECIPHERING DESTRUCTIVE CODE.

AND THESE OPTIC PULSE SWORDS SLASH THROUGH SATURATED BANDWITH.

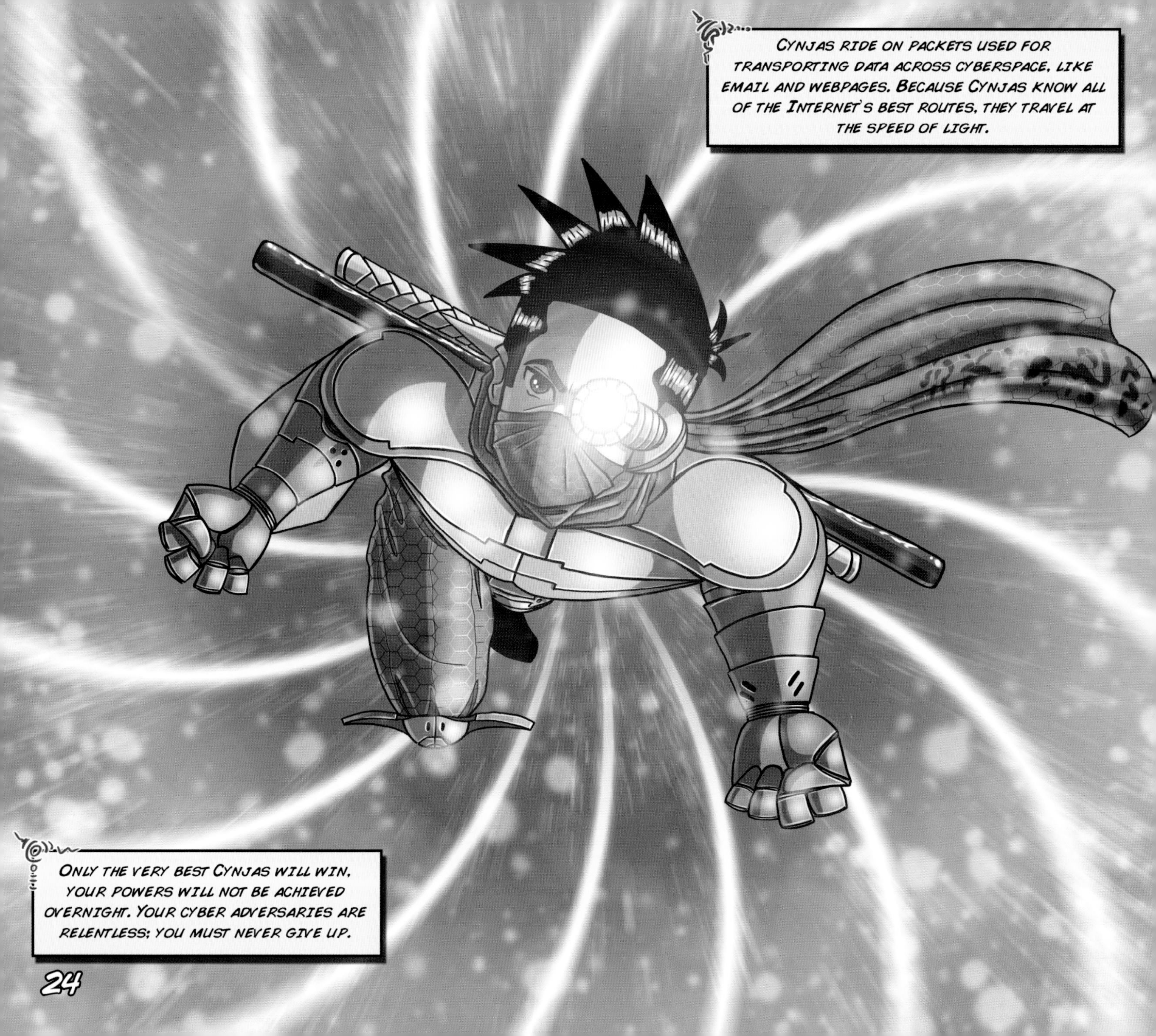
CYNJAS RIDE ON PACKETS USED FOR TRANSPORTING DATA ACROSS CYBERSPACE, LIKE EMAIL AND WEBPAGES. BECAUSE CYNJAS KNOW ALL OF THE INTERNET'S BEST ROUTES, THEY TRAVEL AT THE SPEED OF LIGHT.
ONLY THE VERY BEST CYNJAS WILL WIN. YOUR POWERS WILL NOT BE ACHIEVED OVERNIGHT. YOUR CYBER ADVERSARIES ARE RELENTLESS; YOU MUST NEVER GIVE UP.

Do you accept this challenge?

I will use my powers for protecting all things good in cyberspace.
Yes, Cynsei.

Everything in cyberspace can be used for good or evil. Everyone has a choice to do good or not. Cynjas are noble warriors. They only use their skills for good.

YOUR SKILLS AS A CYNJA ARE ABOUT TO UNDERGO DYNAMIC TESTING.

YOU MUST STAND IN THE PATH OF ZEUS AND EXTINGUISH IT BEFORE THIS TROJAN INFILTRATES COMPUTERS EVERYWHERE.
ONLY YOU CAN STOP THE BOTMASTER FROM WINNING.

UH-OH!
WHAT'S WRONG?
ZOMBIE BOTS ARE BEING ROUTED IN ALL DIRECTIONS.
THEY'RE CLOGGING NETWORKS AND SUCKING FLUIDITY FROM WHAT BANDWIDTH IS LEFT.
YOU JUST FEEL THAT?
SO, WHAT DO I DO ...

... NOW?
ZAP!

GREAT! HE LEFT ME HERE.
GOOD GOING, CYNSEI.

WELL I GUESS IT'S ON ME NOW ...THANKS THERE BIG GUY...
NOT!
SO...IF I STEP INTO THIS PORTAL...

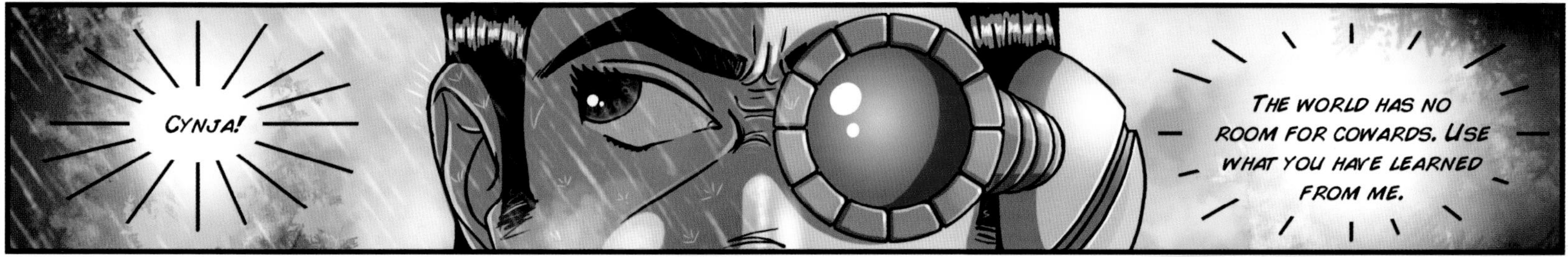
CYNJA!
THE WORLD HAS NO ROOM FOR COWARDS. USE WHAT YOU HAVE LEARNED FROM ME.

OK, CYNSEI...
...I THINK I GOT THIS...

ACROSS CYBERSPACE GOOD COMPUTERS ARE TURNING INTO ZOMBIES AT THE INCREDIBLE RATE OF 3.0 GIGABITS PER SECOND.
STOCK $ EXCHANGE
CARE

WHAT ARE THOSE DUMB ZOMBIES UP TO?

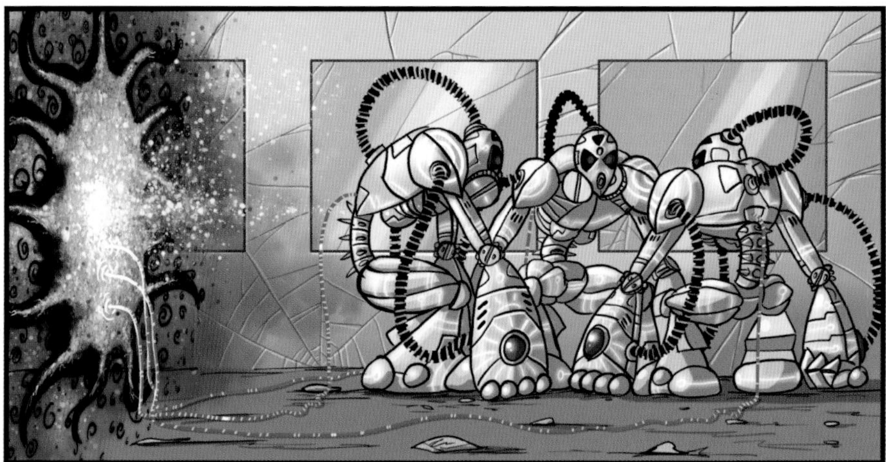

THEY'RE OBEYING THE BOTMASTER'S COMMANDS

IF I MODIFY THEIR CODE, THEN I CAN QUARANTINE THEM.
THAT WAY THEY'LL TURN HARMLESS.

NOW I JUST NEED TO GET THROUGH THAT FIREWALL.

THE CYNJA POWERED HIS BINARY MONOCLE AND STARTED TRACING PACKET ROUTES.

TRACE ROUTE
ACTIVE
PROTOCOL FORCED
THEIR TRAILS SWERVED ACROSS THOUSANDS OF DNS ROUTES, LEAVING BEHIND A RESIDUE OF IP FILTH. FOLLOWING THIS TRAIL OF DESTRUCTION WOULDN'T BE HARD...

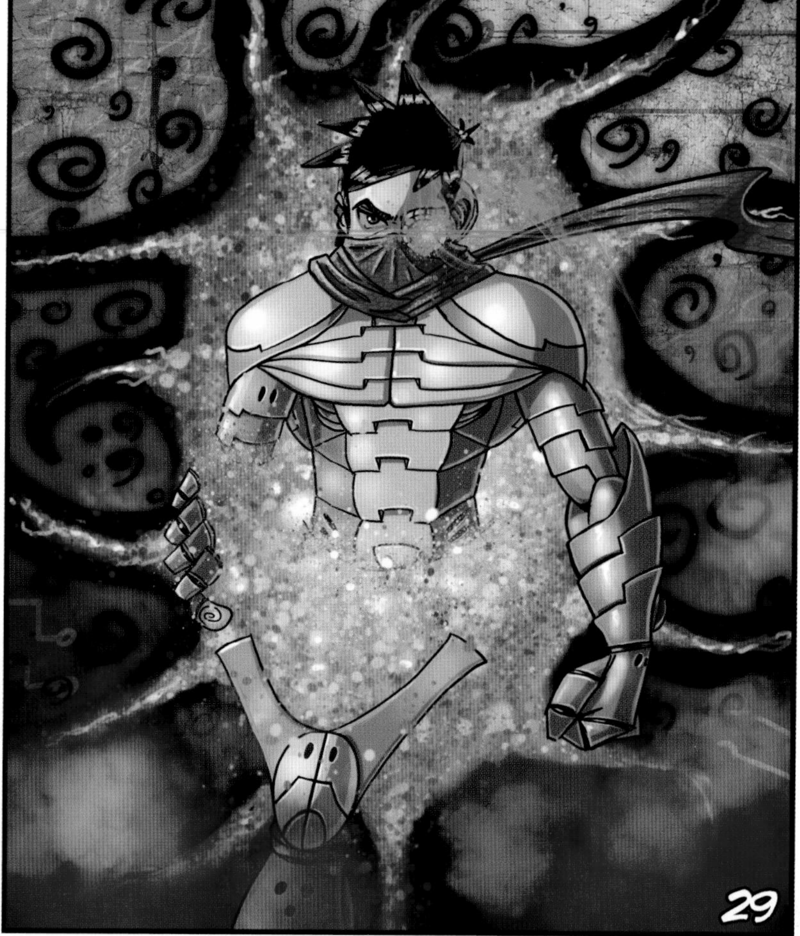

WOW! THIS NETWORK LOOKS CORRUPTED.
BOTMASTER, I KNOW YOU'RE HERE SOMEWHERE!

He zoomed in on an infected file.

These zombies and bots are using anonymous proxies to hide the Botmaster.

There he is.

Where'd he go?

Well hello, little man.

WHOOSH!

UGH!

AAAAAAAAAAAAAAAH!!

PLENTY OF OTHERS HAVE TRIED TO STOP ME.
YOU'RE JUST ANOTHER ONE OR ZERO TO ME LITTLE GUY....

I HOPE YOU LIKE THE DARKNET?
I MADE IT ESPECIALLY FOR YOU.
LET'S SEE YOU CRAWL OUT OF THIS ONE.

Free falling into the Underground

That did not go as planned.

Cynja, what are you doing?!

Cynsei?

I don't have much time, so listen carefully.

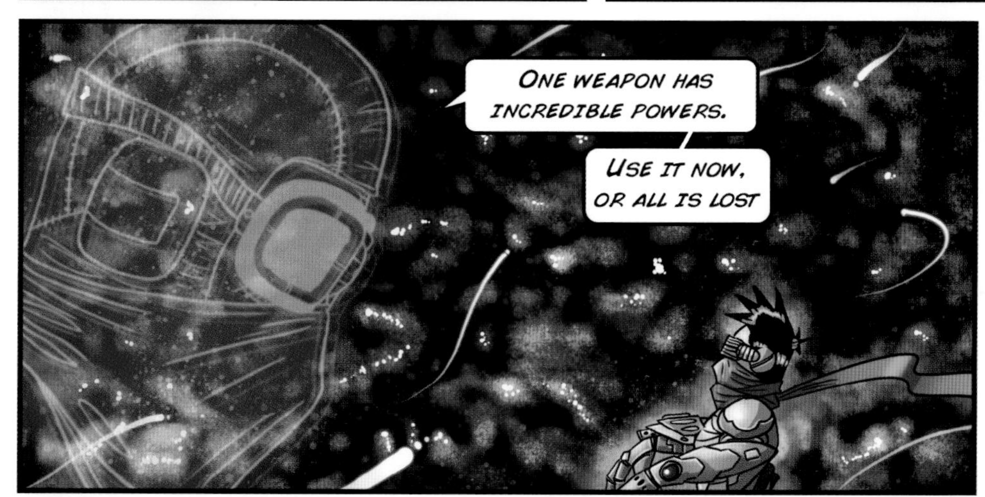
One weapon has incredible powers.
Use it now, or all is lost

Oh great.
He left me again.

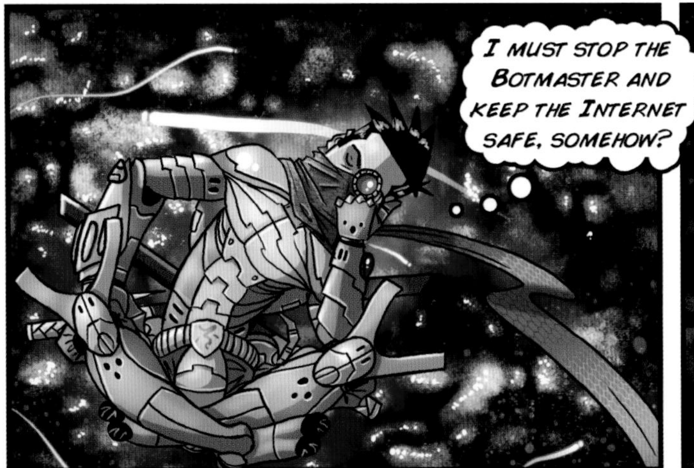
I MUST STOP THE BOTMASTER AND KEEP THE INTERNET SAFE, SOMEHOW?

ZIP!
COME HERE, YOU!

WELL, IF I TAKE BITS OF HEXCODE...

...AND CONVERT IT TO DECIMAL...
FSSSSSHHH...

...THE BOTS WON'T BE ABLE TO COMMUNICATE WITH THEIR EVIL LEADER.
POP!

CIPHER CUBE, SHOW ME THE WAY OUT.
FZZZ
FZZZ

I'M COMING FOR YOU BOTMASTER!!!

ZAP!
CIPHER CUBE. DOWNLOAD CODE.

WHIRRRR

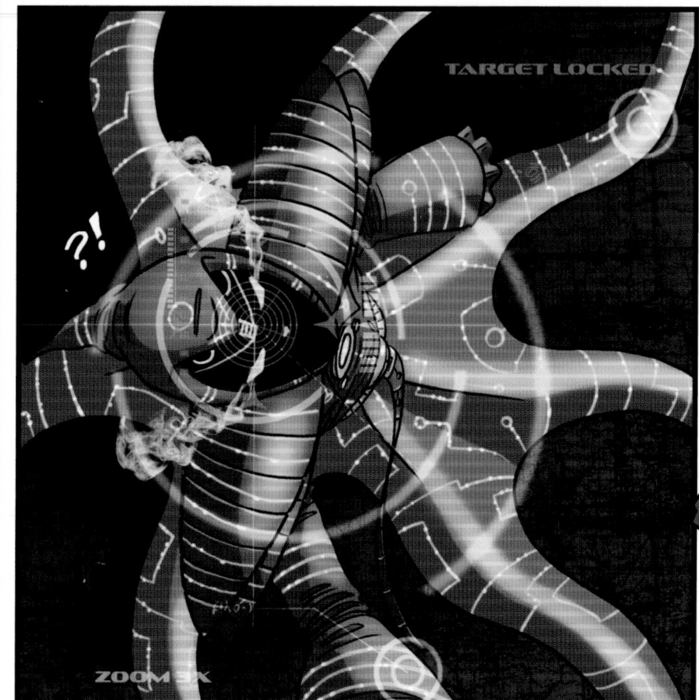
TARGET LOCKED
?!
ZOOM 3X

BOTMASTER! YOU GOT OWNED!
!!
VOOMP
AAAAAAAAH!!
KABOOM

HIS CIPHER CUBE LAUNCHED INTO ACTION.
IT QUICKLY SCANNED FOR VIRAL SIGNATURES AND
QUARANTINED INFECTED FILES. ONE BY ONE
ZOMBIES MUTATED INTO HARMLESS CODE.

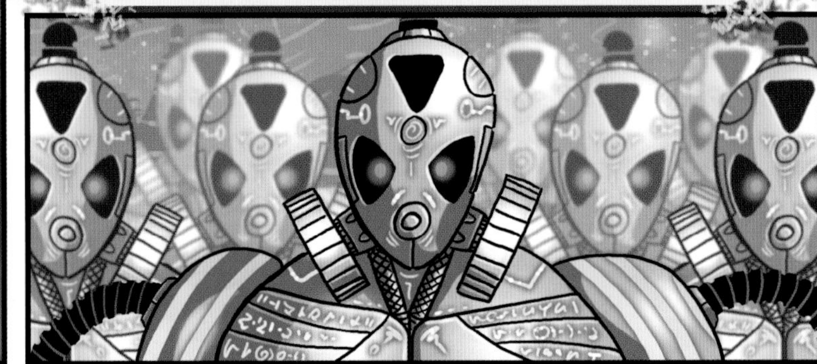

THE CYNJA WAS CERTAIN THE BOTMASTER MET HIS DOOM. HOWEVER, THAT WAS A NAÏVE ASSUMPTION FOR OUR YOUNG CYNJA. AS THE SMOKE CLEARED, HE COULD SEE WHERE THE EVIL LEADER ONCE STOOD.
THE BOTMASTER WAS GONE.
BUT TRACE ROUTE PACKETS FROM HIS BINARY STAFF LINGERED. AT THAT MOMENT THE CYNJA KNEW... THE BOTMASTER ESCAPED HIS FATE UNDER A STORMY HAZE OF CODE.

Cynja!
You must always remember...
... as many times as we win. the bad guys only get better.
Our battles are just beginning!

Bring it on!!

DOMAIN NAME SYSTEM

Its nickname is DNS. This system translates website names into a numerical IP address. An IP address shows where information lives in cyberspace.

WWW.THECYNJA.COM

W0W.TH9CYN7A.CIM

W08.T09CIN7A.CI

208.109.157.41

ENCRYPTION

This is how you make messages secret. Using a special key, information is coded so it looks like nonsense. But if someone else has your key they can decrypt the message and read it.

FIREWALL

It protects against bad guys by controlling the data coming in or going out of a computer network.

INTERNET PROTOCOL ADDRESS

208.109.164.43

208.109.157.41

Every device connecting to the Internet is assigned a unique number called an IP address. This number sometimes shows your location. .

MALWARE

It's short for malicious software that's programmed by bad guys to steal information, send spam, launch attacks or damage your computer. Viruses, worms and Trojan horses are all types of malware.

NETWORKS

Computers exchange information through a complex highway of connections called a network.

PORT

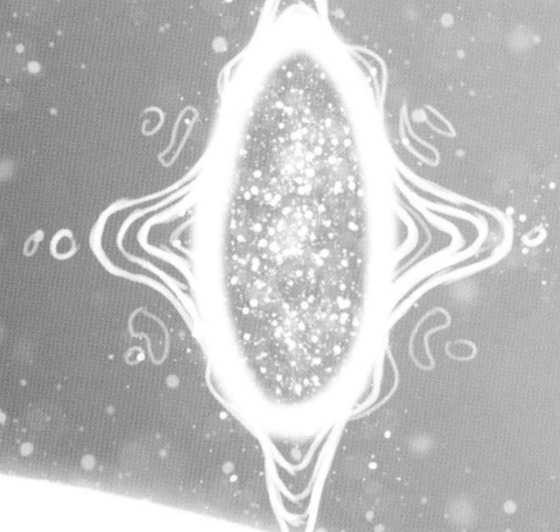

One kind of port is a cable plug-in which connects a computer to another device. Another type helps route data from your computer across the Internet. When a port number is added to your computer's IP address, it tells a server where to deliver the information you've requested while surfing the web or sending emails.

SOURCE CODE

These are rules and instructions your computer must obey.

```
x2=n2;
d_to_b(n2,binary);
for(i=0;binary[i]!='\0';++i)
 bn2[i]=binary[i];
```

VIRAL SIGNATURE

Computer viruses contain a unique fingerprint also known as a code signature. Some malware detection works by scanning files for signatures that belong to known bad guys.

ZOMBIE

When a bad guy hacks a computer and makes it his own, it's called a zombie. Groups of zombies are called botnets. They're used to launch attacks, send spam or steal information. Sometimes you don't even know your computer has turned into one.

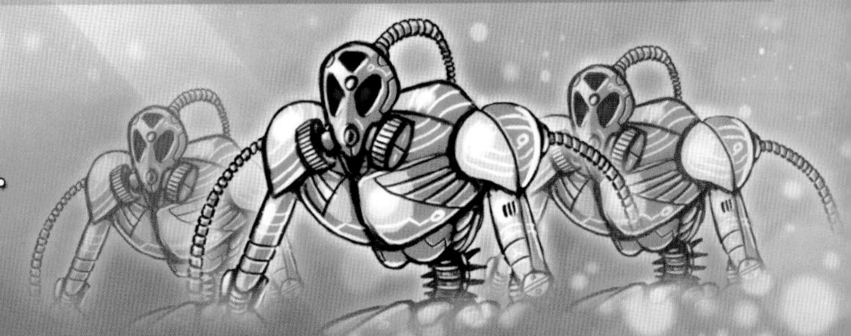

THE CIRCLE OF CYNJAS

VISIT WWW.THECYNJA.COM FOR MORE INFORMATION ON BECOMING PART OF OUR COMMUNITY.